Our Lives Have Gone To The Dogs

Story and photographs by Audrey Spilker Hagar and Eldad Hagar

This book is dedicated to all the animals who are in the shelter right now hoping to be rescued, and in memory of those who never made it out.

Our Lives Have Gone To The Dogs

I've never known life without a dog. When I was born, my parent's Puli, Punim, welcomed me as the newest member of the household by proceeding to lick my face when I came home from the hospital. I, in return opened my mouth in delight. My germaphobic mother managed to restrain herself from separating us. She instinctively knew that this mutual lovefest between the dog and the baby was the start of a bond that would prove us inseparable for the remaining eight years of Punim's life. Punim was never jealous of the attention I took away from her. She accepted my clumsy petting, and tolerated my attempts to dress her up. From the day she met me until her death, she slept by my side. She comforted me throughout the most trying times of my childhood. Punim was my trusted friend and guardian.

At the same time I was growing up in Southern California, Eldad was a young child in Northern Israel with the reputation in his town as the caretaker of the sick and lost animals. Everybody knew to seek him out when they discovered an abandoned bird or motherless kittens. Eldad's parents gave him free reign to bring the animals into his home and help them survive. Eldad's dog Lady, an English Pointer mix, was ten-months-old when she greeted his arrival into the family. It was Lady who helped Eldad with the care and healing of all his orphaned animals. Lady became a mother to all the stray kittens who entered his house. She knew to provide them with the safety and comfort which was so important for their survival. She lovingly accepted the injured hedgehogs and the abandoned baby birds Eldad found as well. Lady died at the late age of seventeen, truly a hero to all who loved her.

Like others who were lucky enough to know such dogs as a Punim or a Lady, we cannot accept the horrible fate that awaits so many dogs in our community. Today in Los Angeles, as a married couple, we are on a mission.

Whether a dog was abandoned and is minutes from being put to sleep at the pound, or is running scared and starved on the streets, we take that animal into our home and become its fosters. We rehabilitate these dogs both physically and mentally by giving them the treatment they deserve. This means that they get lots of love, good food, and live cage and chain free. Finally, we find suitable homes for our fostered dogs, and we begin the process again. As long as the human mistreatment of these magnificent animals continues, we won't stop fighting for them.

The most common question people ask us is, "How can you give them away, don't you become attached?" I love every single dog we have ever fostered and although it's hard to give them up, we remind ourselves that they are going to great homes and that by letting them move on; we now have room to rescue another dog that desperately needs us.

Many of the people who have adopted our dogs keep in touch by sending us photos and e-mail updates. It gives us a lot of joy and comfort knowing we did the right thing. These dogs, who were not just suffering, but were about to be put to death in the shelters, are now happy and loved as they should have been their whole lives. The greatest thing about fostering and adopting dogs is that anybody can do it. Anyone who fosters knows that we are the lucky ones to have such amazing animals sharing our lives, and teaching us the importance of hope, love, loyalty, gratitude, and especially forgiveness.

Time after time they manage to forgive the species whose cruelty and neglect caused them to be abused, used in dog fights, forced out onto the street or dumped in pounds. The power of the dog to not just forgive but to embrace and love a human being is nothing short of miraculous. It may take work and time on the human's side to regain a dog's trust after it has been badly abused, but the dog always forgives.

The second most common question we are asked is, "How can people treat a dog like this?" The only way to reply is with another question, "Why do people do any of the evil things they do?"

It is our job in society to protect those who cannot protect themselves. Human beings created dogs as companion animals and therefore it is our duty as the offspring of those who domesticated and bred these animals to be responsible for their health and well being. The dogs will reward you with renewed spirit and an appreciation for all that the world has to offer. These animals are a gift.

In this book we hope to share our amazing journey with our personal photos and stories of some of the dogs who spent time with us on their road to their second chance at the life they deserved in the first place.

"For every animal that dies in a shelter, there is someone, somewhere, responsible for its death. You cannot do a kindness too soon, because you never know how soon it will be too late."

~Ralph Waldo Emerson

Our Baby Dolly
Found stray when she was only ten-weeks-old

Spotty

Spotty was our first. He was a Pitbull who suffered like nobody should ever suffer. He was a "watchdog", kept in a dirt yard without shelter from the heat or from the cold. He dug a hole to protect himself from the elements the best that he could. Sometimes they fed him. Sometimes they remembered to refill his filthy water bowl. He was lucky because somebody let it be known what was happening in their neighborhood, and Spotty was sprung free.

Spotty was skinny, malnourished, sunburned, and his ears were riddled with fly bites. Because his owners had never adjusted his collar since the time they procured him as a puppy, it had literally grown into his neck and the vet had to remove it surgically. He had fleas, ticks, and internal parasites. His teeth were broken from trying to chew on the metal chain in hopes of freeing himself from the life he was condemned to live. I did not realize then that this scenario would repeat itself so many times. Little did we know that we were just at the beginning of meeting hundreds of animals who had been abused as much or even worse than Spotty.

We met Spotty one week after he had been set free. Though saved from his torturous existence, he was still homeless, living in a temporary cage at a veterinarian's office, recovering mentally and physically. It was our job as volunteers to walk him, give him affection, get him used to being a pet.

The moment the cage opened, Spotty came bounding out and kissed us as if to reassure us that he was going to be okay. All he wanted to do was to lick us and then, when he got the hang of the leash, he just wanted to run because he had never been free to run before and I could swear he was laughing as his tongue hung out of his mouth and flapped in the breeze.

It became our daily ritual. Go see Spotty.

After the warm greeting of kisses, and the crazy run around the neighborhood with the flapping tongue, Spotty would drink out of his pan of water, closing his eyes as if he just tasted the nectar of the gods. Then he would wind down, taking turns sitting in our laps, staring into our eyes, and call me crazy, smiling.

I thought we were going to go help poor Spotty by teaching him a few things. It was he who taught us. We took the plunge, no longer just volunteers walking dogs; we started bringing these animals into our home. We wanted to do as much as we could for all the Spottys who need just a little help.

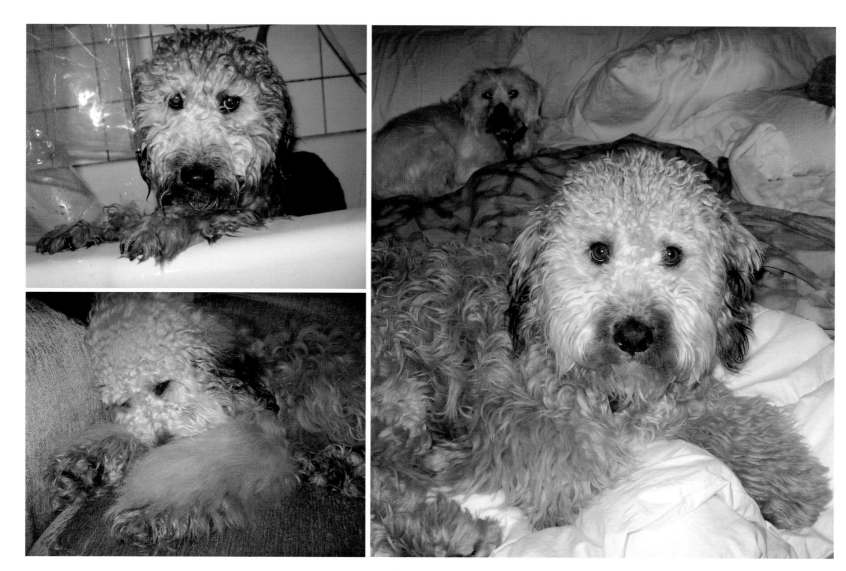

Chester

"Anybody who doesn't know what soap tastes like never washed a dog."

~Franklin P. Jones

"He is your friend, your partner, your defender, your dog. You are his life, his love, his leader. He will be yours, faithful and true, to the last beat of his heart. You owe it to him to be worthy of such devotion."

~unknown

Daisy

Lexi

"God Made the earth, the sky and the water, the moon and the sun. He made man and bird and beast. But He didn't make the dog. He already had one."

~Native American saying

Reggie

Rudy

"Acquiring a dog may be the only opportunity a human ever has to choose a relative."

~Mordecai Siegal

Bambi

Bambi was a petite dog who bore the deep psychological scars of mistreatment by her previous owners. Someone had badly beaten her and she would not let anyone get close to her. The shelter would not allow anyone from the public to adopt her because of her behavioral problems. They would only release her to an experienced rescuer.

We could not touch her without her snapping and crying. When Eldad attempted to pet her, he had to wear an oven mitt to avoid her biting. It was very sad that we needed to teach her that a human hand was not a weapon. When Bambi did bite, she would cry as if she felt bad, as if she didn't want to cause harm but didn't know how else to protect herself.

Bambi was so fearful that she would try to hide and make herself invisible. We once found her sleeping in a shopping bag tucked under a bookcase.

Finally, after two days, she allowed us to pet her head, but she would still snap if we touched her elsewhere. One day when it seemed like there was still no real progress, Eldad decided to go for it by scooping up Bambi and hugging her close to him. She did not cry or try to bite. She tensed up for several seconds, and then slowly relaxed. Bambi changed that day, when she finally felt love and security in the hands of a human.

Bambi's rehab continued as she began to play chasing games and roughhouse with our permanent rescue dog, Dolly. Bambi was now jumping up onto our laps and actually asking for attention. A kind and beautiful woman adopted her and loves her like a daughter. She told us that when she gave Bambi her first bath, she didn't want Bambi to be nervous, so she put on a bikini and jumped into the tub with her.

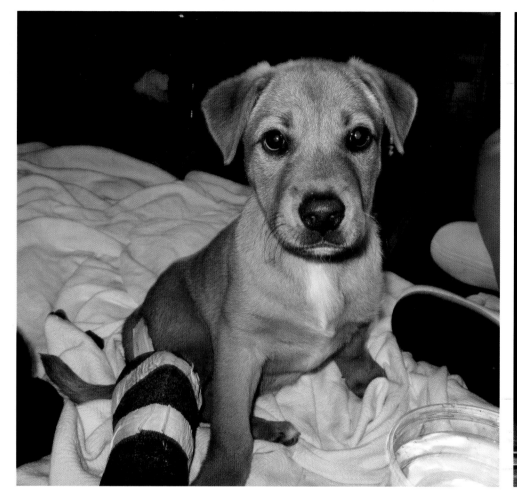

Baby Teddy

Ella

"My little dog - a heartbeat at my feet."

~Edith Wharton

Angela

Andy

"The love of all living creatures is the most noble attribute of man."

~Charles Darwin

Chu-Chi

One of the most common excuses people use for dumping dogs in the shelters is, "We're moving." Chu-Chi, a fluffy beige terrier, was a casualty of such ridiculous reasoning and she was about to be put down. Nobody wanted her because she would back away in her cage when potential adopters would try to pet her. Her time was up. You could read the kindness in her eyes and that she wanted to be touched so badly but was just too scared to allow it. Someone had hurt her before.

Eldad took a chance and put his face up to her cage. She walked up to the bars and gave him a kiss. Since she was already spayed, we were able to take her home that day. She walked out of the kennels on a leash with a sense of joy, her head held high. An L.A. Animal Control Officer even exclaimed, "That dog looks so proud."

She abandoned her fear and turned out to be the kindest and most affectionate dog. A young man who works in the fashion industry adopted her and he gives her a new wardrobe every season.

Our dog Dolly really took a liking to Chu-Chi so she got to have a play date with her at her new home.

"All of the animals except for man know that the principle business of life is to enjoy it."

~ Samuel Butler

Lovey

Popcorn

"I am in favor of animal rights as well as human rights. That is the way of the whole human being."

~Abraham Lincoln

Dolly, Audrey and Mandy

"The opposite of love is not hate, it's indifference."

~Elie Wiesel

Checkers

There was one particular case where I figured we were getting another Bambi on our hands. Again, the shelter was releasing a dog solely to a rescue because the owner who turned her in had described her as a "vicious biter." Eldad agreed to go save this little fluffy terrier mix with the black head and white body.

When Eldad walked in with Checkers, I instinctively moved out of her way. "Just try to pet her," said Eldad. "No way," I said as Checkers came running towards me. "Oh my God," I said to Eldad, "Grab her!" Eldad just laughed at me as I blocked my kneecaps with my hands to fend off this lunatic dog. Checkers didn't attack me. Instead, she licked my hands while wagging her tail.

It turned out that Checkers was one of the happiest and most well adjusted dogs we ever met. Eldad thought it was funny not to mention this piece of information when he set her loose on on me. We tried to make her bite us. We waved fingers in her face and stuck them in her mouth. Nothing worked.

Her previous owners obviously lied. It is common for people to make up wild stories when they abandon their animal at the shelter because they don't want to appear evil or irresponsible for ditching their pet. Of course, their deception endangers the animal even more since the shelter is obligated by law to post a warning; almost guaranteeing the dog's death.

In an odd but fabulous coincidence, Checkers moved into a home with two other black and white terrier mixes whose names are Charlie and Chelsea. The three siblings are excited to celebrate Chanukah together every year.

Heidi

Eldad went to the shelter to bail out a dog when a woman walked in with a small cardboard box. He asked her what was in the box. "Someone threw this into my backyard, I saw them do it and run away." Without looking to see what was in the box, Eldad said, "I'll take it."

Looking back, perhaps Eldad was too quick to accept a small package from a stranger containing a live animal sight unseen. For all he knew, it could have been a baby raccoon, a skunk, or perhaps a relatively large rodent.

As it turned out, it was a starved, little four-pound Maltipoo. She was strangely quiet and calm, yet, a few weeks later somehow morphed into the loudest, boldest dog in our household.

I was afraid someone would stick a tutu on her and carry her in a purse, so we decided that we had to keep her. Dolly and Heidi act as if they have known each other forever and spend hours each day playing and hanging out together. Dolly cleans Heidi's eyes every day and is still waiting patiently for the day that Heidi will reciprocate the favor.

Fifi Tags Heidi

Heidi Steals Corn From A Squirrel

"The reason a dog has so many friends is that he wags his tail instead of his tongue."

~Unknown

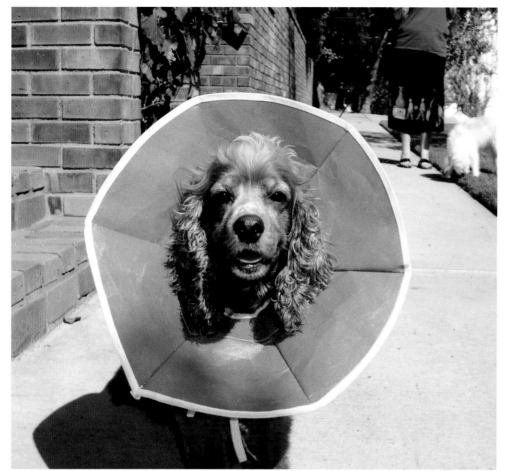

Ruby
(Hurricane Katrina Survivor)

Scooter

"There are two things for which animals are to be envied: they know nothing of future evils, or of what people say about them."

~Voltaire

Junior

Eldad called me and said he had just seen someone heading towards the shelter with a box and intercepted the person before he could turn in the animal. "Was it another Heidi?" I asked. "No," he said, "Even smaller, and I'm bringing it home."

It was not another Heidi; it wasn't even a mammal... it was a baby OWL.

A man had found the owl in the middle of the street in downtown Los Angeles. Eldad rushed the owl to our vet who gave him vitamins and fluids. He displayed signs of shock and appeared very weak. We thought he would die within hours.

We contacted a wildlife rescue and in the meantime, Eldad knew enough about owls to place our newest foster, Junior, in a warm box with a stuffed animal owl in order to make Junior feel like he was back in the nest. Eldad downloaded the sound of owls hooting which he played for Junior. Junior responded by hooting back at the computer.

At nightfall, we peeked into Junior's "nest" and were shocked to see a completely different owl. Junior was standing upright on his stuffed toy and staring at us with clear, intelligent eyes as if to say, "Can I help you?"

While Junior had been resting during daylight, Eldad had procured an owl's feast of mealworms and crickets. I held Junior in my lap and Eldad placed the first worm into the owl's beak. Junior seemed stunned and just let the worm dangle for a second but then he consumed it with great vigor. Junior ate like a king and we knew he was going to survive.

Junior's eyes were the most amazing of all. They were bright yellow and he would stare into the eyes of whichever one of us was speaking, craning his neck to make sure he got a good view. It was extraordinary how intelligent and inquisitive this baby creature was.

Two days later, we took Junior to the California Wildlife Center which is very well equipped to provide wildlife the best chance of survival. The refuge assigned Junior a number that enables us to check on him for as long as he is in their care, and there is a great chance that one day he will be able to care for himself where he belongs; in the wild.

Junior enjoys
a delicious worm

Ralphy

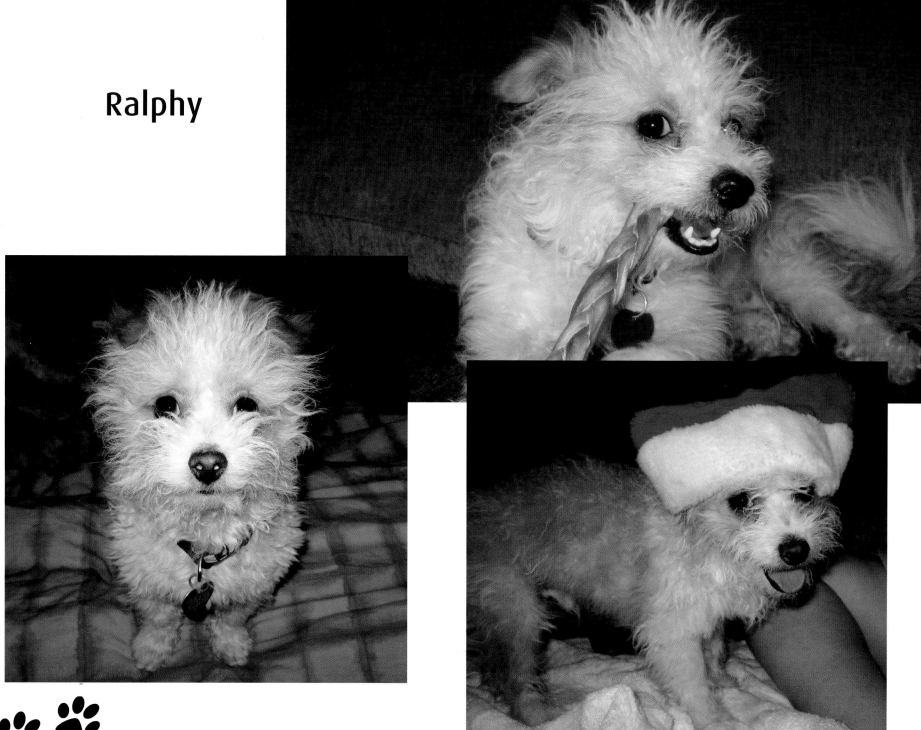

Lady

"The great pleasure of a dog is that you may make a fool of yourself with him and not only will he not scold you, but he will make a fool of himself too."

~Samuel Butler

Fabio

"We can judge the heart of a man by his treatment of animals."

~Immanuel Kant

Muffin
One of Our Rescued Cats

"Thousands of years ago, cats were worshipped as gods. Cats have never forgotten this."

~Anonymous

Prayer of a Stray

Dear God please send me somebody who'll care!
I'm tired of running, I'm sick with despair
My body is aching, it's so racked with pain
And dear God I pray as I run in the rain

That someone will love me and give me a home
A warm cozy bed and a big juicy bone
My last owner tied me all day in the yard
Sometimes with no water and God that was hard!

So I chewed my leash God; and I ran away
To rummage in garbage; and live as a stray
But now God I'm tired; and hungry and cold
And I'm Oh so afraid; that I'll never grow old

They've chased me with sticks; hit me with stones
While I run the streets; just looking for bones
I'm not really bad God; please help if you can
For I have become just another; "victim of man!"

I'm wormy dear God; and I'm ridden with fleas
and all that I ever wanted; was an owner to please
If you find one for me God; I'll try to be good
I won't chew their shoes; and I'll do as I should

I'll love them; protect them; and try to obey
When they tell me to sit; to lie down or to stay!
I don't think I'll make it; too long on my own
Cause I'm getting so weak; and I'm Oh so alone

Each night as I sleep in the bushes I cry
Cause I'm so afraid God; that I'm gonna die
I've got so much love; and devotion to give
That I should be given; a new chance to live

So dear God please; oh please; answer my prayer
and send me to somebody; who will really care
That is dear God; if You're really there!

Anonymous

Taz (Before)

Nobody wanted Taz and time was running out. He sat in the pound covered in motor oil. He was so matted and filthy that the shelter staff accidentally placed him in a female kennel. If he had not been so depressed and had the inclination, he could have gotten several of his kennel mates pregnant.

It is so sad to see that people refuse to acknowlege great dogs because they are dirty or depressed. Who wouldn't be despondent while living for days in a cage after just barely surviving the streets? Who wouldn't be joyless from spending his or her life chained up in a yard without enough food or love?

Taz (After)

Taz started out as a grey dog but a good bath transformed him into a gorgeous young strawberry blonde. His one blue eye and one brown eye added to his unique beauty.

Taz had looks and brains. He was especially good with kids. In less than five minutes, he learned how to shake with one paw and give five with the other.

He now lives in Lake Tahoe where he plays in the snow and swims in the lake.

Bubbles

"If you have men who will exclude any of God's creatures from the shelter of compassion and pity, you will have men who will deal likewise with their fellow men."

~St. Francis of Assisi

Auggie and Muffin

"No one appreciates the very special genius of your conversation as the dog does."

~ Christopher Morley

Jesse, Linde and Brittany

Jesse was a four-year-old black Lab mix. An animal control officer found her lying on the side of the road with two broken legs. She was a victim of a hit and run. This dog needed complex surgery and since the shelters are not equipped to perform such operations, it meant Jesse was to be put to sleep at the end of the day. We could not just call to have her taken off the kill list; someone would have to show up in person to claim her.

Jesse was waiting on borrowed time in "ISO," the isolation unit where they keep the sick and injured dogs. Some animals in this area need operations, while others just have colds and are in quarantine. They happen to be adoptable, but since the public is unaware that they exist, they have slim chances of survival.

Eldad came to her rescue just as the doors were shutting for the night. Then, as he carefully carried Jesse out to the car, a woman ran to the door holding a kitten she had just found on the street. Eldad told her that the cat would immediately be killed by the shelter because it would require special care at such a young age. The girl started crying and explained she was from out of the country and was soon leaving to go back home so she could not keep the three-week-old kitten. Eldad then took the kitten from the woman after he had placed Jesse comfortably in the backseat of his car.

He rushed Jesse to the veterinarian, since she would be receiving early morning surgery. The orphaned kitten, we named Linde, came home to stay with us. Neither of our two adult female cats initially showed any affection toward her. Because Linde was tough and had a huge appetite, it seemed likely that this orphan would pull through. Eventually, our cat Mousie, who likes to clean all the dogs who pass through our house, finally decided to groom Linde.

41

Jesse came home the next afternoon, still groggy with two bandaged, repaired femurs. She was incapable of standing on her own and we had to carry her outside when she had to go to the bathroom. Though recovering from mental and physical trauma, the instant Jesse the dog met Linde the kitten, she decided to become her new mother. There was a little competition between Mousie and Jesse for the affection of little Linde who was very happy ambushing all the animals and biting their tails.

Jesse began to heal and finally walk on her own. The bond she shared with Linde was now unbreakable.

We received a great application for Jesse from a very special family willing to give her the extra medical care she needed. We crossed our fingers and mentioned how she happened to have a best friend who was a kitten. Unbelievably, the husband who was hoping to adopt Jesse confided that he was a cat lover and would be more than happy to adopt little Linde along with Jesse. It was a dream come true for everyone. Linde was renamed Meowzer by the family's eight-year-old son. It was miraculous how everything came together.

Custody Battle

The story did not end with the family who had adopted Jesse and the newly christened Meowzer. It just so happened that when we went for our routine house check of the family's home, we were accompanied by our newest foster Brittany who was a gorgeous and rare Polish Lowland Sheepdog who had been left at the shelter because her family said she grew larger than they expected.

Brittany was a fluffy ball of shaggy cuteness, full of fun and energy. The wife's mother could not resist her charms. She wanted Brittany and since we loved this family so much, we agreed. We had placed two dogs and one cat in one week; we thought we hit the jackpot.

A week later, we received a call from the family concerning Brittany. The father sadly said that he would have to give Brittany back to us. He explained, "Basically, there was something that had me worried," he continued. "Our son was playing in the backyard on Sunday with some of the kids from church and well, Brittany got very excited and decided that she needed to herd all the children so she was chasing them and pulling their pants down."

We found Brittany a new home without children and today she shares a yard with a Golden Retriever who taught her how to swim in the pool.

Kiss me you fool!

Brittany

"I think dogs are the most amazing creatures; they give unconditional love. For me they are the role model for being alive."

~Gilda Radner

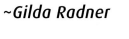

Kylie on Maggie

What kind of dog are you?

"No matter how little money and how few possesions you own, having a dog makes you rich."

~ Louis Sabin

49

Casey, Pepper and Gigi

"...he will be our friend for always and always and always."

~ Rudyard Kipling

Dasha

There is nothing better than seeing a dog who once was in terrible shape, completely transform into the epitome of beauty and happiness. Dasha was an American Bulldog who was obviously starved. The vet discovered that she had uterine tumors that had to be taken out right away. She was so skinny and weak we did not know if she would even survive. Her emaciated appearance was so shocking that Eldad was afraid to walk her in public because he thought people would mistaken him as a dog abuser. Dasha pulled through her operation and after a short recovery, a loving couple drove eight hours from San Francisco to adopt and take her home.

Six months later, Dasha and her parents drove back down to Los Angeles in hopes of possibly adopting another dog. We were excited to see how Dasha had turned out. Obviously, we expected her to be healthier and hopefully a little heavier, but we were surprised to see that Dasha was actually chubby. This former skeletal stray now eats like a queen and has a top-notch veterinarian who has recommended that she lose about five pounds.

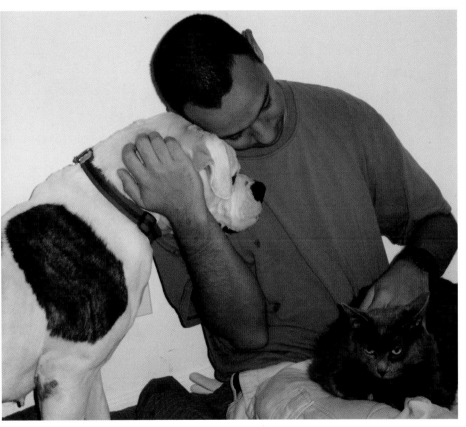

"To err is human, to forgive, canine"

~Anonymous

Sammy

"Dogs are miracles with paws."

~Susan Ariel Rainbow Kennedy

Zoey

The animals who come into the shelters as strays off the street have the "luxury" of a few days to live at the shelter because of the fact that they may actually be lost and have owners who are searching for them. However, the animals whose owners intentionally dispose of them have very little time to live because nobody is looking for them. Since so many dogs are packed into the shelters, the odds of someone adopting them are stacked against them. Some people turn in their dogs when they think they are too old and want a puppy instead; others will discard their pet if it gets sick or has fleas. We rescued a purebred Shih-Tzu named Zoey whose pound papers said she was given up due to "snoring." She was so cute and mellow and of course has a great home now and her owner loves her so much.

Chloe

"There is no psychiatrist in the world like a puppy licking your face."

~Ben Williams

Rusty

Rusty was turned into the shelter and was too downhearted to even raise his head to acknowledge the people who came to his cage. It was as if he had given up on life. He also had a major bout of kennel cough, an ailment that is common amongst shelter dogs or wherever many dogs are confined together, yet is easily treatable. Rusty was a Jack Russell Terrier, an active breed that cannot stand to be pent up. Eldad said he never saw such a dejected and miserable dog and therefore he had to get him out of the pound.

Rusty definitely had personality. He was irritable from being ill and cooped up. His old owners mistreated him so he would cower whenever we picked up an object. He just wanted to be left alone. However, just two days of TLC brought out Rusty's happy side. As he started to feel better, he revealed to us that he had an obsession; playing fetch. Now, the majority of dogs love to play with toys or at least they learn to love to play with toys. A somber fact is that so many of the dogs we rescue have led such joyless lives that they do not understand the concept of playing. They have to be introduced for the first time to toys and then taught that it is okay to chase and chew on them.

Whether or not Rusty had ever seen tennis balls before, he immediately grew to have an extreme love for them. All Rusty wanted was for us to throw the ball so he could catch it and he made us throw it over and over and over again. He was fast and if we were not fast enough for him or ignored him, he would drop the ball on our feet. If we didn't immediately throw the ball, he would bark and beg us to hurry up.

We thought that Rusty had stored energy and frustration from living in a cage and hoped that a day at the dog park would tire him out. At the dog park, he was quicker on his feet than every dog of every size. He chased every ball that we threw. He stole other dogs' balls, then approached strangers and dropped those stolen balls onto their feet. He was insatiable.

When we finally left the park after several hours, he fell asleep in the car. We smiled smugly; we had worn him out. Once we got home, Rusty suddenly had a renewed sense of vigor, found his tennis ball, stared me in the eye, and proceeded to drop it on my foot. It was then that we noticed BLOOD! Rusty had run himself ragged and his paws were chapped and bleeding. This dog was so obsessed that even self-injury could not sway his unwavering need to chase the ball.

We were at the point of considering getting Rusty some psychiatric treatment when we got a call from a man who trained dogs to perform at science fairs in a humane show aimed at teaching people the natural talents of dogs. He explained that this entire troop was composed of rescued dogs. He felt that Rusty was totally focused but not dominant towards other dogs and this meant he'd be a perfect fit.

We met the man, his wife and the troop and were amazed at the beauty and grace that his pack of assorted rescues displayed in their act. These happy, well-adjusted dogs loved to perform as a means to harness their energy. This was not an animal circus, these were just dogs acting like dogs. They were constantly rewarded, yet never forced to perform. The troop encouraged rescuing and adopting dogs.

Rusty passed his audition and was adopted. They wanted him even if it turned out that performing was not in his genes. They saw him as a new family member regardless of his talents. As it turned out, Rusty was a natural and today is famous across the country as the renamed, infamous "Action Jackson." Rusty can finally use his speed and agility and his beloved ball in a focused and positive manner instead of being manic and frustrated. His keen concentration and agility has made him a star. He is also very much loved and that is the most important part of his story.

"Dogs come when they're called; cats take a message and get back to you later."

~ Mary Bly

Goliath

Occasionally we get a call for help with an unusual animal. One such wild animal we took care of was Goliath, the Sugar Glider. Sugar Gliders are marsupials which are native to Australia and New Guinea. They are small creatures so people think they are cute pocket pet alternatives to rats or hamsters but what they don't realize is that these little animals are much more complex than pet rodents and need an enormous amount of care and diligence when it comes to the intricacies involving their feeding and habitat. Unfortunately, people come to realize that they take a lot of work and consequently the poor animals end up not cared for properly as well as being passed around from home to home.

Goliath was traumatized. We were the third family we knew of who had taken him in. All he wanted to do was hide in his little nesting pocket. He only emerged at night to eat. Normal Sugar Gliders need a balanced diet of protein and calcium. Goliath was never fed the correct diet and had been deprived of medical care.

Sugar Gliders have extra flaps of skin that give them the ability to float off tree branches. They are social creatures who need a lot of space to climb and dive. They are actually happiest when they are high up. Poor Goliath had been living in a flat hamster cage on the floor in an apartment where noisy kids would bother his daytime sleep.

Eldad built Goliath a huge enclosure with stairs and toys and ramps made of eucalyptus from which he could dive. He also had a special area high up where he could sleep undisturbed in his pouch. During the days he was with us, we kept noise to a minimum until the sun went down and Goliath woke up.

Because Goliath required special conditions, we could not just let anyone adopt him. We were fortunate enough to find Goliath a permanent home with a woman who owns a wildlife rescue. This organization houses wild animals that people bought both legally and illegally, thinking it would be fun to own an exotic animal. However, they soon realized that wild animals need to live in the wild instead of under a roof or chained in a backyard. Sadly, since these animals are now domesticated, they lack the skills to survive in the wild so they have to live out the rest of their days in these sanctuaries.

Beatrice

Some dogs seem to take forever to find a home. One of these was a Rottweiler named Beatrice. Nobody wanted her simply because most people are fearful of adopting certain breeds. Just as the shelters are full of cast off Pit Bulls, the same holds true with Rotties who do not make it out alive.

Beatrice did have a commanding presence that drove other dogs crazy, but she was as gentle as could be and was especially good with children. On the way to adoption events, she demanded that we allow her to sit in the front seat. We had to tie her leash to the headrest or else she would try to sit in Eldad's lap while he was driving.

One day at a huge adoption event sponsored by Animal Planet, a mother and her five-year-old daughter stopped by our booth and took an interest in Beatrice. I was holding Beatrice's leash as the daughter was petting her head and I was describing Beatrice's great personality to the mother. The intelligent little girl was telling me all she knew about Rottweilers.

Just as the daughter was kissing Beatrice goodbye, a Dalmatian walked by and stared at Beatrice just as all dogs tended to do when they caught sight of her. Beatrice usually ignored other dogs but this time she stared back at the Dalmatian. Everything seemed to happen in slow motion. I looked down and saw Beatrice's leash in my hand but Beatrice was no longer attached to it. She was running full force at the Dalmatian. Time seemed to stand still as Beatrice was charging toward the other dog and the little girl and her mother were watching the whole scenario. I couldn't breathe. Beatrice stopped right in front of the Dalmatian and started wagging her behind. She acted as if she was in love with him. She had merely wanted to stop and say "Hello Gorgeous!" The Dalmatian wagged back and then Beatrice trotted back to our booth.

An Animal Planet employee, who already owned a Rottie, fell in love with Beatrice. We had been hoping to find Beatrice a home with another dog.

Bessy

Bessy came to us covered with ticks because of being left outside and neglected. She was an older dog afflicted with Pyometra, a potentially fatal disease that only affects unspayed dogs. She was bailed out of the shelter seconds before she was supposed to be put down. Our vet, Dr. Werber, performed a life saving hysterectomy on Bessy and she quickly recovered. Mentally, Bessy was still a mess. She didn't know how to play with toys, she didn't like cats and she didn't like dogs. Since she was victim of maltreatment, she also was afraid of most people. We feared that nobody would want to adopt her.

Unexpectedly, a friend of ours called and said that she was hoping to find a dog for her mother-in-law. We took Bessy to visit the mother-in-law in the Hollywood Hills. The woman spoke very loudly. "What's her name?" she asked. "Bessy," I said. "Betty," answered the woman. "No, Bessy," said Eldad. "Oh, Betsy," she exclaimed, "Hello Betsy, come here Betsy!" Bessy hid behind Eldad at first, but then slowly approached the woman who immediately asked us if she could keep her. We agreed and figured it would take Bessy a few days to adjust. We had to distract Bessy and make a run out the door when she was not looking. No one would ever guess, especially Bessy, that this was a match made in Heaven.

Bessy's story was out of a movie. She went from existing in a backyard to living the good life in a plush house where she spent her days relaxing with her loving new owner, watching television, and being hand fed by the woman's two caretakers.

Bessy became very attached and protective of her mistress. When the woman past away a little more than a year later, we received a phone call from our friend who was sobbing and said, "Bessy has gotten under the covers with my mother-in-law and she won't let the paramedics take her."

Now Bessy was heartbroken. Just when she found happiness, it was taken away and she needed a new home. Luckily Bessy was adopted right away. The love she received had changed her, and she in turn had brought joy to the woman during her final days. The woman's son says that Bessy gave meaning to his mother's life.

Even more astonishing is that the dog who did not have social skills, who was treated like garbage for most of her life, and was thought unfit to even live her life, now goes to work everyday at a hospital as a therapy dog for the elderly.

Misty, Sydney, Caesar, Phoebe and Daphne.

Misty, a tiny terrier, came to us sick, hungry, and sad. She ignored Dolly and the cats. Two weeks later, she gained a few pounds, recovered from kennel cough, and ultimately came out of her shell.

Soon after Misty showed up, we got a call that a dog and her three puppies were at high risk of euthanization. Eldad went to the shelter and brought the four of them home. The young, reluctant mother, Sydney, preferred to spend as little time as possible with her offspring. We had to hold her down every two hours, and bribe her with treats just so she would feed Daphne, Phoebe, and Caesar.

Caesar and Phoebe seemed healthy, but Daphne, the runt, seemed to be failing. She was incapable of opening her eyes while the other puppies were thriving. We braced ourselves for the worst.

Meanwhile, Misty divided her time between playing chase with Dolly and Sydney and fostering Sydney's puppies. Misty took it upon herself to groom the three pups and even attempted to nurse them. This care probably is what ultimately saved Daphne's life. Misty then found the perfect home.

Although Sydney and each of the puppies found separate homes, we have stayed in touch with their new families and every year we celebrate the puppies' birthday with their mother, Sydney. They just turned three on April 18, 2008.

We cannot tell if they know that they are related, but they seem to have a lot of fun.

Daphne, the former sickly runt is the largest one of all. In fact, her owners just recently adopted another dog from us, Baxter, the sweetest Cockapoo we saved from the East Valley shelter just as his owner was getting rid of him. They are the cutest brother and sister who just love each other.

Topsy

Topsy looked like a Muppet. Her new owners live right next door to a three-legged Chihuahua we rescued a few years ago called Mimi. They meet every day at the beach in Malibu. Topsy is a healer of sorts. When her owner's old dog had past away, their pet Cockatiel, Lolly, became very sad. Lolly mourned her dog friend by laying eggs. This seemed to put her in a worse temper. Once Topsy entered her life, Lolly was happy again and she finally stopped laying eggs now that she has a new canine sister who is also her best friend.

"A human being is a part of the whole called by us universe. Our task must be to widen our circle of compassion to embrace all living creatures and the whole of nature in its beauty."

~Albert Einstein

74

Mindy

"To be followed home by a stray dog is a sign of impending wealth."

~Chinese Proverb

Baxter, Niko, Bibi and Dixie

"The greatness of a nation and its moral progress can be judged by the way its animals are treated...I hold that the more helpless a creature, the more entitled it is to protection by man from the cruelty of man."

~Mahatma Gandhi

Rex

Rex was the cutest white terrier who needed surgery to fix his leg. He was very gentle but extremely sad and could barely walk. Our first priority was to raise the funds to get him the medical attention he needed. The longer he waited the less chance he had to ever walk normally.

The stars were in alignment when an amazing couple offered to adopt Rex into their family. We explained Rex's medical situation. Their response was, "Go get him the surgery, we'll pay for all of it, no matter how much it costs."

A few days later, Rex underwent the repair of his leg. Dr. Olds, his orthopedic surgeon, inserted a metal pin to aid in its proper healing. Coincidentally, his new owner had been injured and had a pin in his arm as well. Rex's pin was eventually removed and he now goes on long hikes and runs faster than all the dogs in the park. Thanks to his adoptive parents, both his physical and emotional pain are ancient history.

Witnessing the incredible transformation of dogs like Rex, and meeting caring and generous people such as his adoptive parents, is what keeps us motivated and inspired.

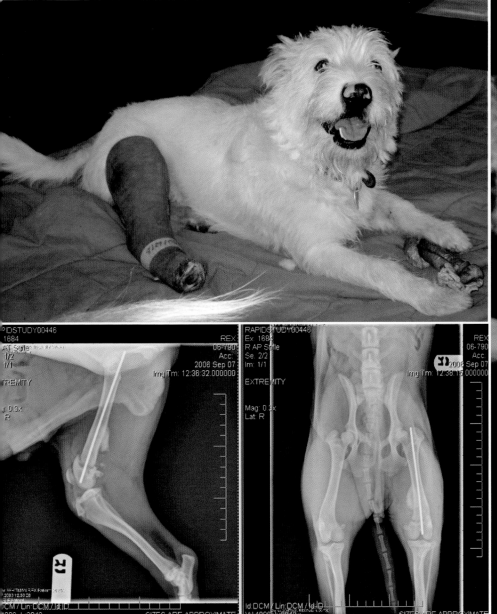

Bindy and Quincy

Bindy and Quincy were dumped at the shelter together. The inseparable canines were terrified. It is not easy to find an adopter who will agree to take in two dogs together. However, we were determined not to break them apart; no matter how long it might take. It was worth the wait when a wonderful couple decided they loved Bindy and Quincy at first sight and wanted them both to live in their lovely home.

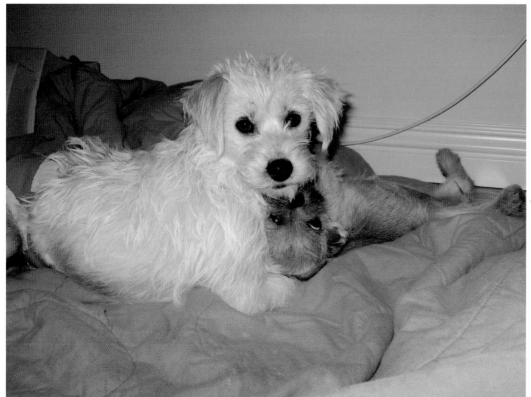

Bindy and Maddy

"Scratch a dog and you'll find a permanent job."

~ Franklin P Jones

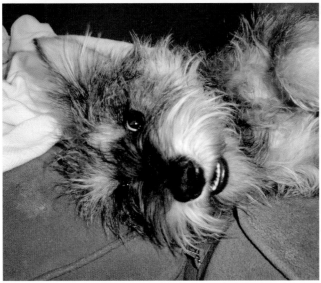

Rosie

Rosie was a one-year-old American Bulldog with a herniated left eye that was sightless. She was picked up as a stray and the dogs in her crowded kennel had attacked her. When we got her home that morning, she crept along the ground as if she was afraid of being attacked again. By nightfall, she was completely paralyzed in all four of her limbs.

Fearing that she had a brain tumor, we took her to the vet the next day. We readied ourselves for the worst. The doctor ran several tests, but nothing conclusive came back explaining what was wrong with her. Rosie was about to be sent to the neurologist when the doctor started feeling around in her armpits. "Aha," he exclaimed. We looked down to see a couple of ticks that were hiding deep within her sockets. The vet said, "Rosie has a case of Tick Paralysis." The ticks had been embedded there for so long, they had injected large amounts of their paralytic toxin into her body.

Throughout her ordeal, Rosie was easy going and upbeat. She slowly began to walk again. As soon as she was strong enough, her bad eye was removed. She and Dolly became best friends. Dolly tried to help her with her physical therapy by walking her on the leash.

We wondered who would want to adopt a limping, one-eyed American Bulldog. We weren't kept in limbo for long at all. Rosie soon found a wonderful woman to adopt her.

Today she goes for long runs on her private beach with her mom and is happy as a clam.

She stays in touch with Dolly and stops by when she's on the West Side.

Photo by: L.A. Animal Services

Roxy

One evening, someone dropped his pet dog Roxy at the shelter. The horrible stress of the abandonment and the shelter environment took its toll. She crashed before we could even get to her, and she passed away. No matter how cute you think your dog is and that you believe someone will adopt it, you are risking its life in so many ways.

"If there are no dogs in heaven, then when I die I want to go where they went."

~Unknown

Daisy and Minnie

Daisy was not spayed and her irresponsible owner allowed her to become pregnant when she was practically a puppy herself. She was almost the same age as her daughter Minnie who was tossed alongside her at the shelter by their owner.

We miraculously found them a home with another fabulous family willing to take the two of them.

Lizzie and Toby

Lizzie and Toby had never met before they came to our house from separate shelters. Somone hit Lizzie with their car and didn't stop. A kind citizen saw Lizzie lying on the side of the road and tried to help her. Although this person could not afford to pay for medical attention, she had the good sense to take Lizzie to the closest shelter where they could provide her with some care. Eldad was in that particular shelter and saw her in the medical room which is hidden from the public. He immediately brought her home. Lizzie had a black eye, a bleeding nose, and a sprained leg. Lizzie was terrified by everyone. If someone walked towards her, she backed away.

Toby, a Dorkie, (a Dachshund and Yorkshire Terrier mix) and his sister were dumped because their owners were having a baby and felt that it would be inappropriate to actually keep their own dogs with a newborn in the house. This was really a blessing in disguise as we could clearly see that Toby had been maltreated by these owners. He was way too thin and was afraid of being touched by both humans and dogs. Somebody adopted his sister out of the shelter, but poor Toby was left behind. To make matters worse, another dog in his cage kept harassing him. Toby was such a sad mess and he had contracted kennel cough.

Fast forward two weeks: with a little love and kindness, Toby became the funniest clown of a dog who jumped on the bed and flipped over for belly rubs. His little short legs made him look like a sea otter on its back. Meanwhile, Lizzie's eye cleared up and the limp went away. She became the friendliest most adoring animal who ever lived. She wanted to kiss every person, cat, and dog... especially Toby.

Lizzie became Toby's new sister and luck was smiling on them when a great family adopted both of them. We are thrilled they will be together forever.

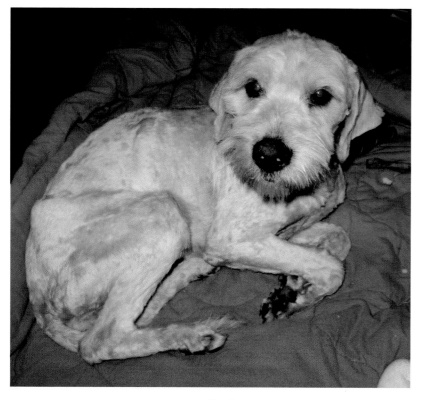

Teddy

Hours after we got him out of the shelter

Teddy

A few days later, feeling MUCH better

"Until one has loved an animal, a part of one's soul remains unawakened"

~Anatole France

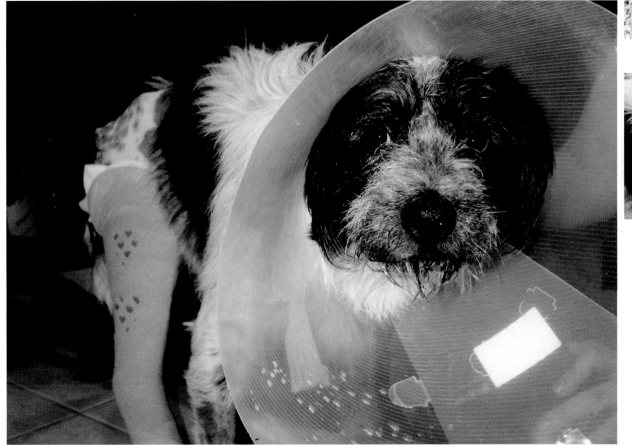

Champ
(Before & After)

Foxy

Frenchie

Missy

Mousie And Billy

AFTERWORD

When we began photographing our fostered dogs, we never imagined we would write a book to commemorate them; we just wanted to remember the dogs we loved. Then our friends and family encouraged us to get the word out to others: to educate the public about animal abuse and neglect; and inspire people to adopt and rescue animals in need. We have been so moved by these animals who shared a small part of their lives with us, we decided to create our own non-profit animal rescue organization, **Hope for Paws.**

It takes such little effort on our part as humans to make a difference. It is unbelievable what big changes one can make with a few small acts of kindness. You can do it too.

This book illustrates how much joy you can derive from sharing your life with an animal which had at one time been disregarded and considered worthless. So many dogs in the shelters have been overlooked by many people because they showed shyness or fear. Or else, many an animal's filthy appearance or flea infestation, were a turn off to many a prospective owner (all easily remedied situations) resulting in people missing out on a fabulous lifelong companion and ultimately the tragic death of the unwanted pet.

We can tell you from our experience that to ignore these animals is a huge mistake. These dogs are not throwaways. They go on to become search and rescue dogs, assisted living dogs, therapy dogs and more.

A dog you save today, might one day save your life or at the very least, be your best friend.

For more information on how you can help an animal in need please visit our website at:

www.hopeforpaws.org

Special Thanks To:

• Kari Whitman and all the volunteers of Ace of Hearts Dog Rescue
• The officers and ACT's at the Los Angeles City and County Shelters
• Dr. Jeffrey Werber, Dr. Robert Olds, and Dr. Dean Graulich with their
  devoted and remarkable teams at their respective facilities:
  Century Veterinary Group, Brentwood Veterinary Clinic and Malibu
  Coast Animal Hospital.

And a huge THANK YOU to all the animal rescuers and volunteers all
around the world.